OUR SOL

MARS

by Alissa Thielges

AMICUS

dirt

volcano

Look for these words and pictures as you read.

crater

rover

Is that a red star?

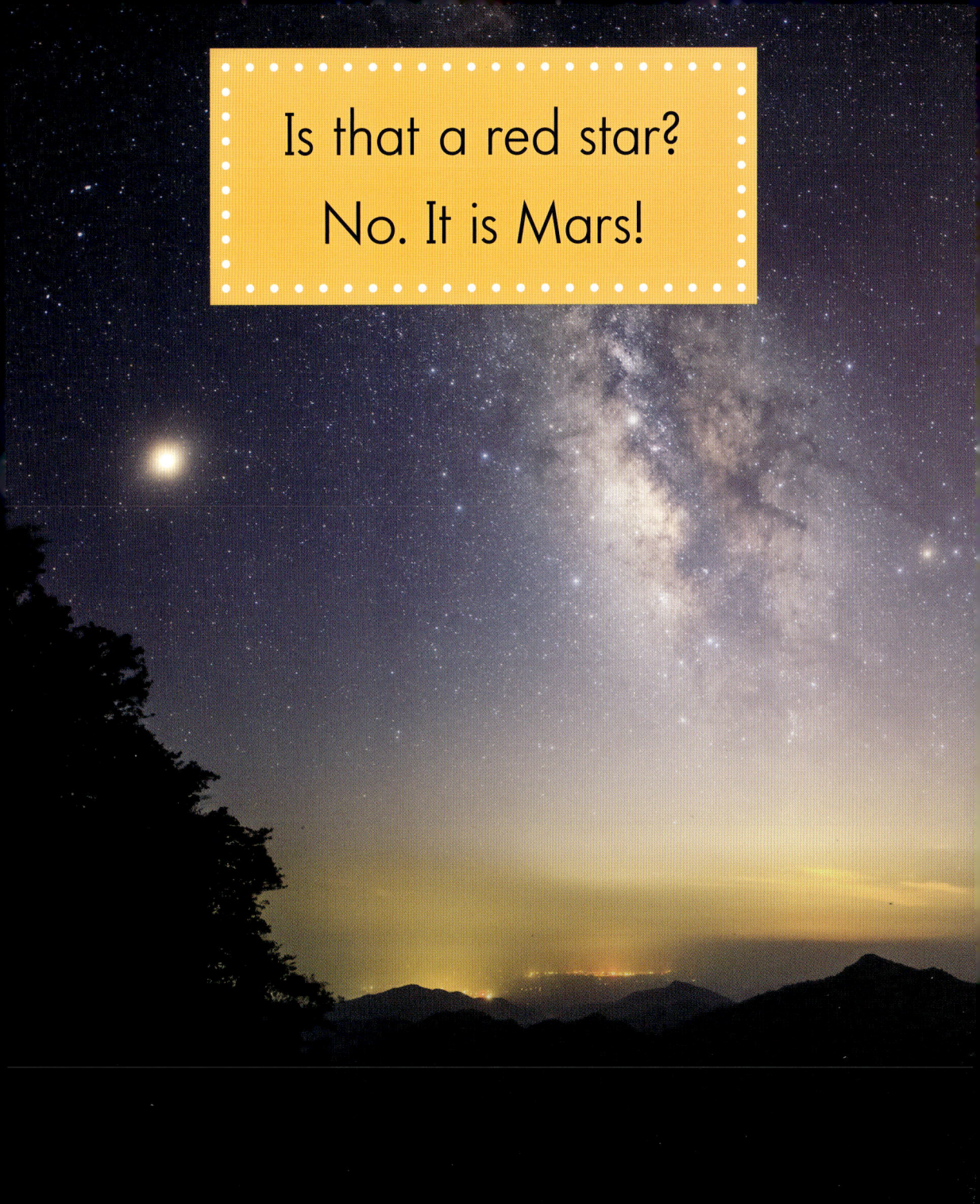

No. It is Mars!

Mars is a planet.
It goes around the sun.
It is the fourth planet from the sun.

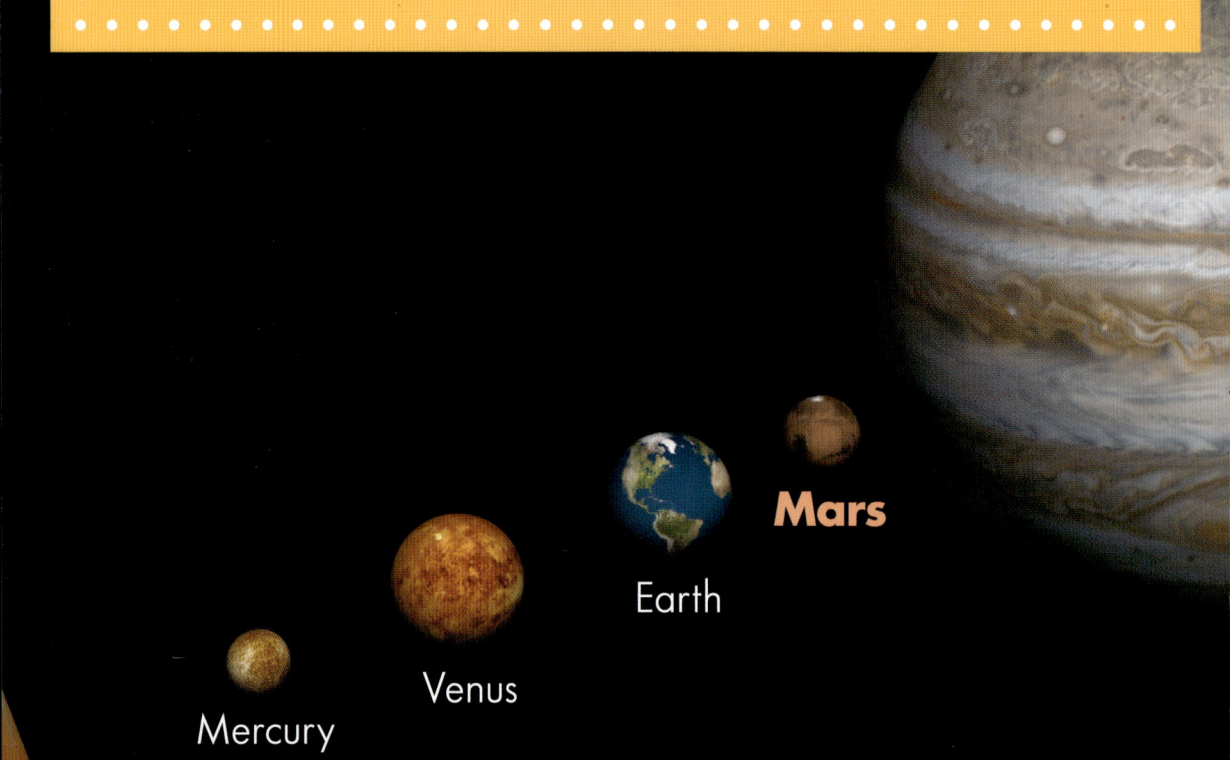

Sun

Mercury

Venus

Earth

Mars

Jupiter

Saturn

Uranus

Neptune

See the dirt?
It has iron. Iron rusts.
That is why the dirt is red.

dirt

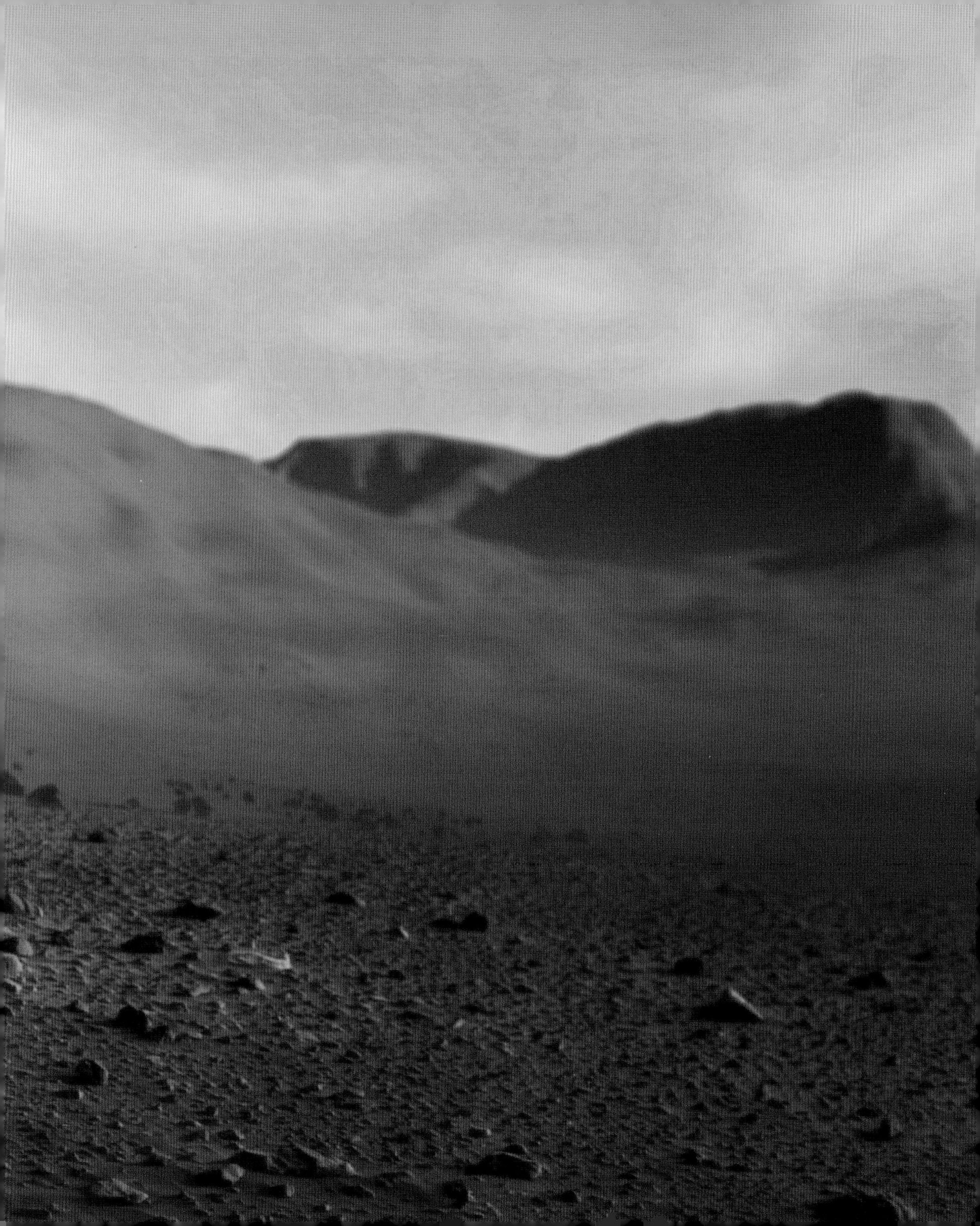

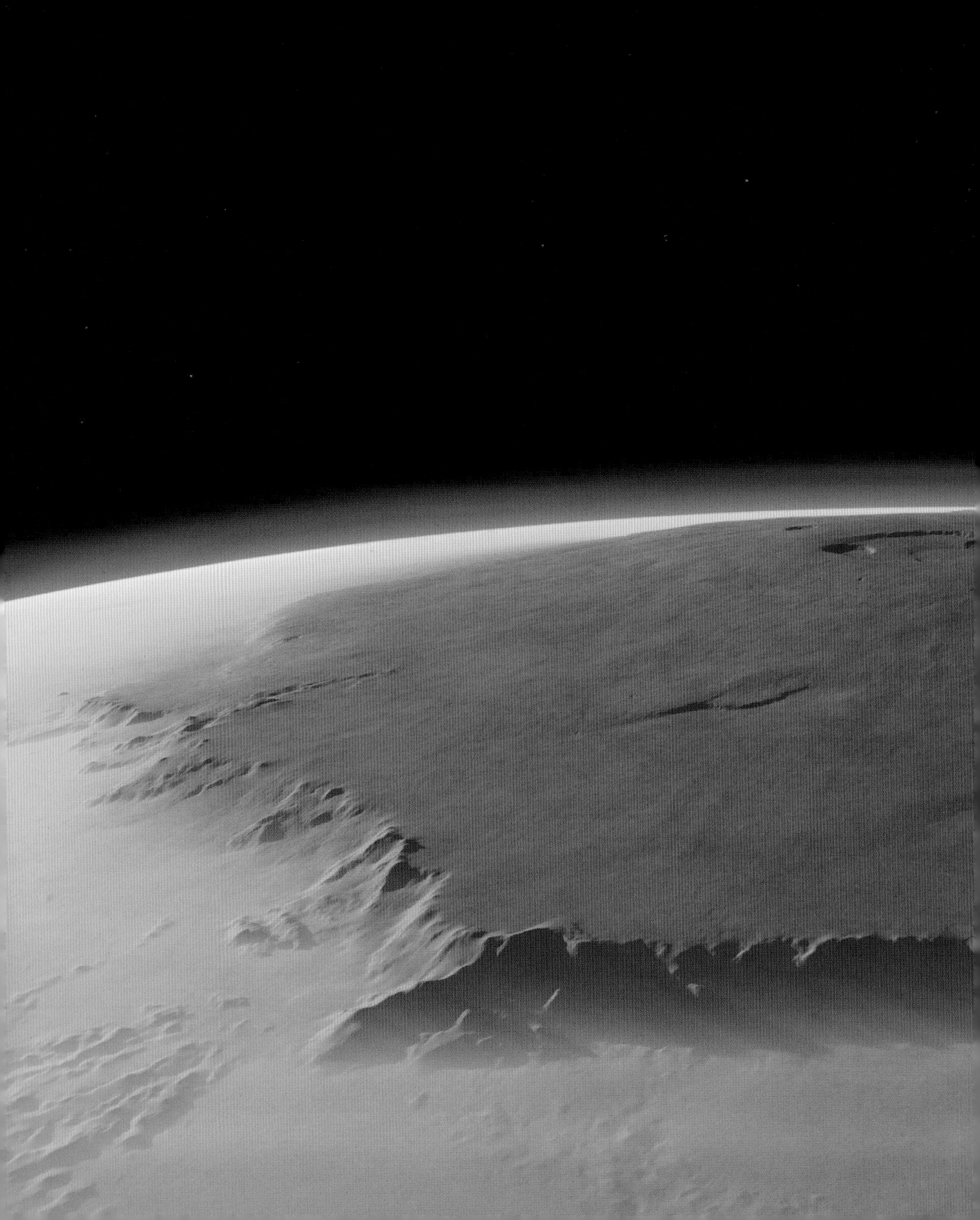

See the volcano?
It is named Olympus Mons.
It is the tallest in the solar system.

volcano

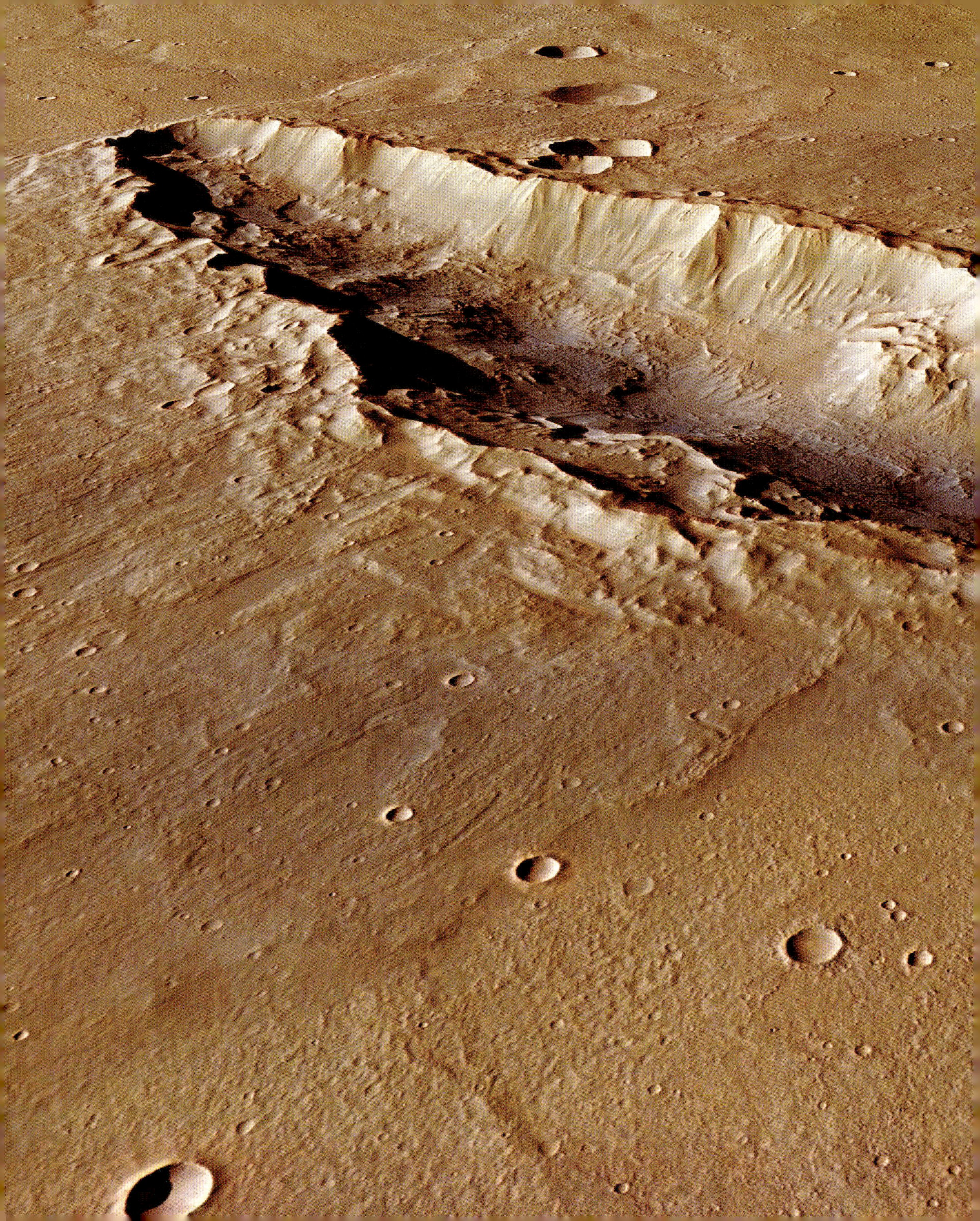

crater

See the crater?
A big rock came from space.
It hit Mars. Bam! It left a crater.

See the rover? It is on Mars.
It collects rocks and dirt.
These are studied back on Earth.

rover

People study Mars. Could we live there someday? Maybe!

dirt

See the dirt?
It has iron. Iron rusts.
That is why the dirt is red.

dirt

volcano

See the volcano?
It is named Olympus Mons.
It is the tallest in the solar system.

volcano

Did you find?

crater

See the crater?
A big rock came from space.
It hit Mars. Bam! It left a crater.

crater

rover

See the rover? It is on Mars.
It collects rocks and dirt.
These are studied back on Earth.

rover

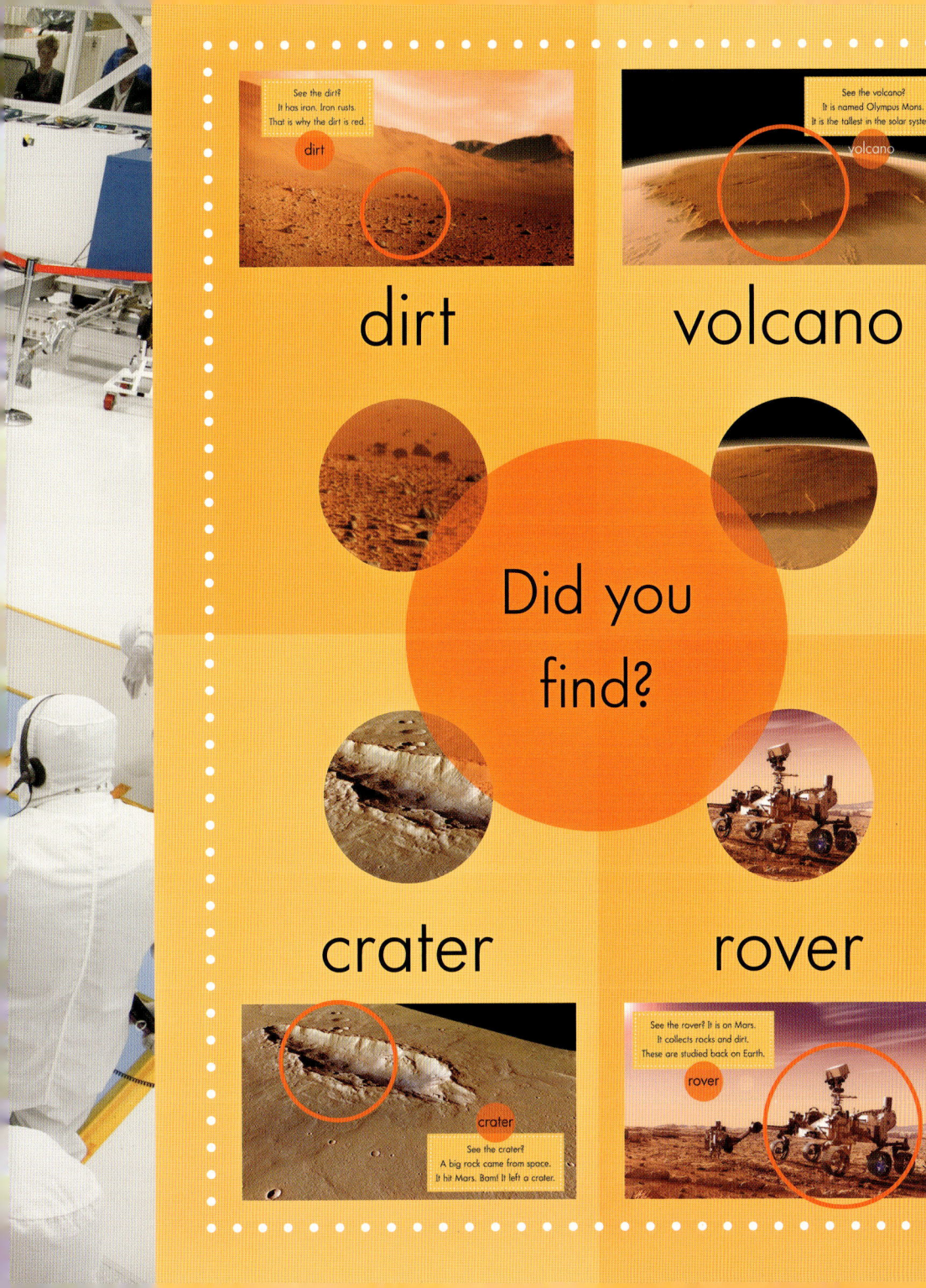

Spot is published by Amicus Learning, an imprint of Amicus
P.O. Box 227, Mankato, MN 56002
www.amicuspublishing.us

Copyright © 2024 Amicus. International copyright reserved in all countries. No part of this book may be reproduced in any form without written permission from the publisher.

Library of Congress Cataloging-in-Publication Data
Names: Thielges, Alissa, 1995- author.
Title: Mars / by Alissa Thielges.
Other titles: Spot. Our Solar System.
Description: Mankato, MN : Amicus, [2024] | Series: Spot. Our Solar System | Audience: Ages 4-7 | Audience: Grades K-1 | Summary: "Simple text and a search-and-find feature reinforce new science vocabulary about Mars' geography and NASA's latest land rover for early readers"—Provided by publisher.
Identifiers: LCCN 2022035877 (print) | LCCN 2022035878 (ebook) | ISBN 9781645492672 (library binding) | ISBN 9781681527918 (paperback) | ISBN 9781645493556 (ebook)
Subjects: LCSH: Mars (Planet)—Juvenile literature.
Classification: LCC QB641 .T45 2024 (print) | LCC QB641 (ebook) | DDC 523.43—dc23/eng20230106
LC record available at https://lccn.loc.gov/2022035877
LC ebook record available at https://lccn.loc.gov/2022035878

Printed in China

Rebecca Glaser, editor
Deb Miner, series designer
Lori Bye, book designer
Omay Ayres, photo researcher

Photos by Alamy/Stocktrek Images, Inc. 8-9; ESA/DLR/FU Berlin/G. Neukum 10-11; Getty/ewg3D 4-5, MARK GARLICK/SCIENCE PHOTO LIBRARY 6-7; NASA/JPL-Caltech/Mars Science Laboratory/14NASA/NASA/JPL/MSSS cover, 16; Shutterstock/Artsiom P 12-13, carlosramos1946 1; TWAN/Jeff Dai/3

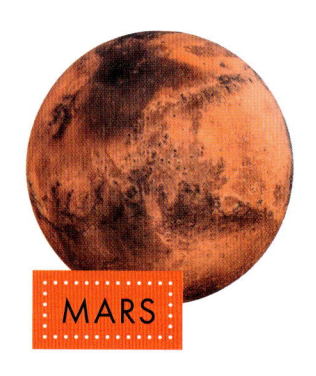